Barefoot Books

INCREDIBLE ANIMALS

written by **Dunia Rahwan**

illustrated by **Paola Formica**

Barefoot Books
step inside a story

Table of Contents

As you travel through the darkest jungles, hottest deserts, deepest oceans and highest mountains of the world, meet the fascinating animals in each category below.

Top Predators

A **food web** is a diagram that shows the connections between two main types of animals. **Predators** hunt and eat other animals. The animals that get eaten are called **prey**. Top predators, also called **alpha** or **apex predators**, are the best hunters because they are smarter, stronger or larger than all the others. They are at the top of the food web because they have no predators themselves. However, humans can still be a threat to them.

Cheetah

The cheetah's special skill is its speed. Cheetahs can weigh 55 kg (125 lbs), as much as an adult human. But when they're hunting, a cheetah can run up to speeds of 112 km/hr (70 miles/hr), which is as fast as a car might drive on a main road. Cheetahs live in Africa and Asia, where they have to watch out for other strong predators such as lions, hyenas and leopards. Animals like these often try to steal a cheetah's prey!

Orca

Orcas, also called killer whales, are top predators because of how intelligent they are. They live and hunt in social groups called "pods." When they're hunting, the orcas in a pod work together so cleverly that few creatures can escape their attack. Mother orcas teach their hunting techniques to their children. They can even change their hunting strategies based on the prey and the environment. Orcas can live in all oceans, but they are most commonly found in colder waters.

Harpy Eagle

The harpy eagle has an enormous **wingspan** of 2 m (6.5 ft) — wider than most humans are tall! You might think this would make it hard for the bird to fly through the dense **rainforests** of Central and South America, where it lives. But its excellent eyesight helps it navigate while chasing prey. The harpy eagle likes to hunt sloths and monkeys, which it attacks with its extremely sharp **talons**, or claws.

Komodo Dragon

The Komodo dragon is not really a dragon, but at 3 m (10 ft) long and 100 kg (220 lbs), it is the largest lizard in the world. Found in Indonesia, it is also the only **reptile** top predator that lives on land. Komodo dragons hunt deer, pigs and water buffalo. They can kill their prey with one **venomous** bite.

Saltwater Crocodile

Any creature should be scared if they find themselves swimming near a saltwater crocodile! The largest reptile on the planet, the crocodile is about as heavy as a small car. It can grow to be 5.5 m (18 ft) long, almost the length of a school bus. The crocodile also has the strongest bite of any animal on Earth. It lives in southeast Asia and northern Australia.

Wolverine

The wolverine might look like a small bear, but it is actually a member of the weasel family. Wolverines are so ferocious, they can capture prey up to five times as large as themselves. They live in the far northern areas of North America, Europe and Asia. In this cold **habitat**, they hunt animals that can easily get stuck in heavy snow, such as reindeer, sheep and moose.

Wolf

Wolves travel and hunt in social groups, called "packs," of three to 30 wolves. Wolves in a pack cooperate to chase and capture their prey. They change their strategies depending on what type of animal they're hunting. For example, wolves are not as strong as a moose, but they can run for longer. The pack will follow a moose until it is too tired to keep going and then attack. Wolves live in similar areas to the wolverine.

Great White Shark

The great white shark is the largest predator fish in the world. It changes its hunting techniques depending on where it finds its prey. It hunts animals as large as sea lions, sea otters and dolphins. An adult female great white shark can weigh more than 3,000 kg (6,600 lbs), about as much as a pickup truck. Despite their enormous size, these sharks can swim more than 30 km/hr (18 mi/hr) while hunting and can leap out of the water to catch prey. They live in coastal areas of all the oceans.

Brainy Beasts

Humans are not the only intelligent **species**, or type of animal. Scientists have observed many other animals showing signs of intelligence such as making tools, communicating, learning from experience, planning future actions — and even showing kindness to one another!

Chimpanzee

The chimpanzee is a **primate**, the category of **mammals** that includes monkeys and humans. Chimpanzees are able to solve problems about as well as a three-year-old human child. They live in Africa, where they teach their young how to heal diseases by making medicines from plants and how to find food using tools available in nature. For example, chimps will use stones and twigs like a hammer and nail to break open nuts.

Portia Spider

This jumping spider is a type of **arachnid**, a creature with eight jointed legs. The portia spider can play tricks! It hunts insects and other spiders in rainforests in Australia and southeast Asia. Sometimes it climbs into another spider's web and pretends to be a captured insect. Then it will attack and eat the spider that owns the web when it comes over to investigate. The portia spider also sneaks up on insects from behind, and will sometimes eat other jumping spiders.

New Caledonian Crow

The New Caledonian crow can make and use tools to find food, unlike any other bird. The crow can hold a twig steady in its beak, then insert the twig into cracks in tree trunks or holes in the ground to go "fishing" for insects to eat. It lives on the islands of New Caledonia in the Pacific Ocean.

Bottlenose Dolphin

Of all the types of dolphins, the bottlenose is the most intelligent. They can solve problems quickly and efficiently. They can even learn from their own mistakes and successes. As well as all this, bottlenose dolphins have shown empathy, the ability to care about each other. If one dolphin is injured, other members of its group will protect it from predators and hold it up to the surface of the water to help it breathe. They live in warm ocean waters all around the world.

African Elephant

Female elephants are known for their incredible memories, which they use to help the rest of their herd find food and watering holes. The **matriarchs**, or female elders of a herd, can also tell whether a far-off elephant is a friend or an enemy, just from the sound of their trumpeting. They are found in rainforests and deserts in Africa.

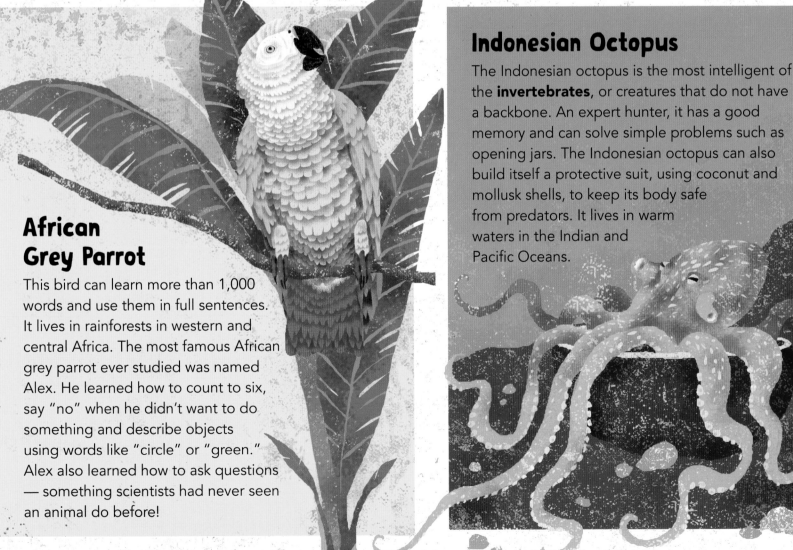

African Grey Parrot

This bird can learn more than 1,000 words and use them in full sentences. It lives in rainforests in western and central Africa. The most famous African grey parrot ever studied was named Alex. He learned how to count to six, say "no" when he didn't want to do something and describe objects using words like "circle" or "green." Alex also learned how to ask questions — something scientists had never seen an animal do before!

Indonesian Octopus

The Indonesian octopus is the most intelligent of the **invertebrates**, or creatures that do not have a backbone. An expert hunter, it has a good memory and can solve simple problems such as opening jars. The Indonesian octopus can also build itself a protective suit, using coconut and mollusk shells, to keep its body safe from predators. It lives in warm waters in the Indian and Pacific Oceans.

Small But Deadly

These creatures are tiny — the largest is less than 20 cm (8 in) long!
But they can all kill their prey with a single bite, sting or punch.
Be careful not to annoy these creatures, because some of
them are deadly to humans.

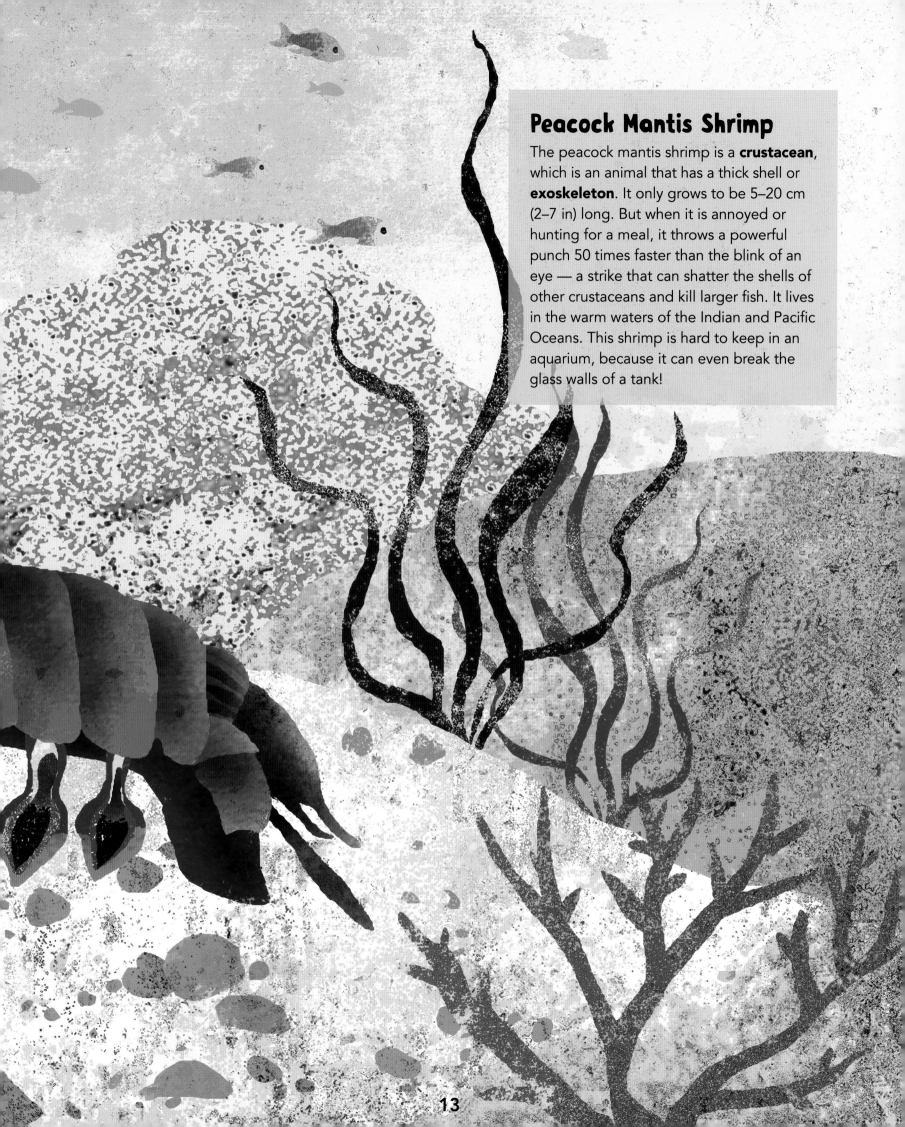

Peacock Mantis Shrimp

The peacock mantis shrimp is a **crustacean**, which is an animal that has a thick shell or **exoskeleton**. It only grows to be 5–20 cm (2–7 in) long. But when it is annoyed or hunting for a meal, it throws a powerful punch 50 times faster than the blink of an eye — a strike that can shatter the shells of other crustaceans and kill larger fish. It lives in the warm waters of the Indian and Pacific Oceans. This shrimp is hard to keep in an aquarium, because it can even break the glass walls of a tank!

Giant Desert Centipede

At 20 cm (less than 8 in) long, the giant desert centipede is only about the length of a regular dinner fork — but it is one of the most dangerous centipedes in the world. It lives in the deserts of North America, where it hunts insects, lizards, frogs and even rodents. Two long pincers behind its legs pierce its prey and inject a powerful venom.

Bullet Ant

This tiny insect lives in the rainforests of Central and South America, where it hunts other insects. Even though it is only 3 cm (less than 2 inches) long and its venom is not deadly to humans, the tremendous pain from its bite can last up to 24 hours.

Indian Red Scorpion

The Indian red scorpion can only grow up to 8 cm (3 in) long, but it is one of the most dangerous scorpions in the world. It can kill humans with a single sting, although it will only sting people to protect itself. It lives in India, Pakistan, Nepal and Sri Lanka, where it is **nocturnal**, only hunting at night.

Cone Snail

This creature only grows about as long as a pencil. It hunts fish, mollusks, worms and other snails that live in the Indian and Pacific Oceans. The cone snail has a long **proboscis**, a tube that extends from its mouth. This tube shoots out a harpoon-like tooth with a venomous sting. Once the snail has paralyzed its prey, it uses its proboscis to drag the meal into its mouth.

Irukandji Jellyfish

This is the smallest jellyfish in the world . . . and one of the deadliest! Its body, called the "bell," is only about the size of a marble. The bell and **tentacles** are covered in venomous stingers. This poison instantly paralyzes prey such as small fish. Irukandji jellyfish live off the coast of Australia. They are so delicate that they cannot survive in an aquarium, which makes them difficult to study.

Blue-Ringed Octopus

The blue-ringed octopus is one of the most venomous animals in the entire Pacific Ocean. When it feels threatened, bright blue rings appear all over its body. It may be only the size of a golf ball, but its bite has enough venom to kill 26 adult humans within minutes. The blue-ringed octopus hunts small fish, crustaceans and mollusks. First, it bites the prey and injects it with venom. Then it grabs the animal with its arms and eats it.

Clever Camouflage

Some animals can **camouflage**, or copy the appearance of other animals or plants in their environment. They do this to hide from predators or to sneak up on prey!

Orchid Mantis

This small insect camouflages itself to look like the petals of an orchid that grows in the rainforests of Indonesia, Malaysia and Sumatra. It sits quietly on the flower, blending right in, until another insect lands on the orchid to drink its nectar. Then the mantis traps its prey using its powerful legs covered in sharp spines.

Banded Snake Eel

This eel is harmless, but it looks very similar to the yellow-lipped sea krait, a venomous snake that also lives in the Indian and Pacific Oceans. The eel also copies the way the sea krait moves. When other animals try to hunt it, they are tricked into thinking it's the dangerous sea snake and they swim away without attacking it.

Decorator Crab

This crustacean does not have a protective shell like other crabs do. In order to survive, it has learned to disguise itself. It chooses pieces of **coral**, **anemone** or sea sponge and arranges them on its back. They stay attached thanks to a special fuzzy covering that works like Velcro. Decorator crabs can be found in oceans all over the world.

Leopard

A leopard's spots help it camouflage with its environment. Leopards can be found almost anywhere in the world and the patterns on their coats look different depending on where they live. Leopards that live in dimly lit areas beneath tall trees have lots of darker spots to help them blend in with the shadows.

Leaf-Tailed Gecko

This gecko only lives on the island of Madagascar. While hunting for prey or hiding from predators, it flattens itself against tree bark and blends in so well, you'd have to be an expert to spot one! The gecko's tail is shaped like a leaf, which makes it even better at camouflaging.

Rock Ptarmigan

This bird lives in the **tundra**, or very cold northern areas of the globe. During the winter, it has thick white feathers, which keep it warm and help it disguise itself in with snow. But in the summer, it loses its white feathers and grows brown ones instead, so it can hide among branches and in muddy meadows when the snow melts.

Cuttlefish

In less than one second, the cuttlefish can camouflage its appearance to blend in with its surroundings. Special **cells** in its body let it change its skin to any shade of the rainbow. The cuttlefish lives in warm, shallow waters worldwide, although it has been found in deeper waters, too. It often hides by shrinking its body and moving its tentacles to look like **algae** swaying in the water.

Brilliant Beaks

These birds all have beaks that have **evolved**, or changed, into strange shapes! Evolution means that an animal species changes gradually over many generations. These changes help the animals survive better in the environments where they live.

Pelican

The pelican uses its beak as a net to catch fish. The lower part of its beak has a stretchy "throat pouch" that can hold over 11 L (3 gal) of water. That's as much water as you could fit in a kitchen sink! The pelican catches fish in this part of its beak. Then it spits out the water and swallows the fish. Pelicans live along coastlines in many parts of the world.

Rhinoceros Hornbill

This bird looks like it has two beaks stuck together. The top piece, which looks like a large horn, is called a casque. It acts like a megaphone, making the bird's call loud enough to be heard throughout the southeast Asian rainforests where it lives. This enormous beak might look heavy, but it is actually very light. It's made of **keratin** — the same material as human fingernails!

Sword-Billed Hummingbird

This tiny hummingbird's beak is longer than its body. It only weighs 10–15 g (less than half an ounce), as much as a dozen paper clips. It uses its sword-like beak to reach the nectar of long, thin flowers found in the Andes mountains of South America. While it drinks, the hummingbird beats its wings constantly. This allows it to hover in the air next to the flower.

Roseate Spoonbill

This bird's beak looks like a pair of large spoons! It lives in the southern United States, where it hunts in swamps for small fish, crustaceans, algae and insects. It sticks its beak into the muddy bottom of the swamp and moves its head back and forth, using its beak like a spoon to stir up food.

Shoebill

The shoebill is a type of stork, a long-legged bird that wades in shallow water. It lives in Africa, where it hunts fish, frogs and even baby crocodiles. Its powerful beak is shaped like a shoe. The shoebill hides in bushes or plants until it sees prey, before it sneaks up and grabs the animal with its strong beak.

Toco Toucan

This bird's beak is one-third of its body size. However, this enormous beak is actually very light. It's made of keratin and bone, and filled with many pockets of air. With such a big beak, the toucan can reach fruit on long branches and dig out insects from holes in trees. It lives in warm areas of northern South America.

Black Skimmer

This bird lives along coastlines in North America. The bottom part of its beak is longer than the top. This shape seems to help the skimmer when it's hunting fish. It glides along just above the water, with its beak skimming the surface. When it finds a fish, it closes its beak very quickly, trapping the prey.

Pink Flamingo

The flamingo's beak is a water filter! This large pink bird lives near shallow, muddy beaches. It turns its head upside down in order to dip its beak completely underwater. Then it swings its head side to side and uses its tongue to push water through its beak. The beak is lined with rows of tiny plates, like a comb. These plates allow dirt and water to flow back out, trapping algae, insects and other food for the flamingo to eat.

Amazing Migrations

The animals in this section all **migrate**, or travel very long distances
from their homes. Some of them migrate in enormous groups that
can include millions of animals! Most migration happens when
seasons change or when it's time for the animal to have their babies.

Gnu

The gnu is a type of antelope that migrates between Kenya and Tanzania each year. Millions of gnu travel together, searching for new sources of food and water during the rainy season. Thousands of zebras, gazelles and impalas migrate around the same time. In such large groups, each animal is more protected from predators.

Grey Whale

This whale migrates further than any other mammal. Scientists have tracked seven grey whales that swam from Russia to Mexico and back. One of them, a whale they named Varvara, swam more than 22,000 km (13,670 mi) in 172 days. That's like swimming back and forth across the entire United States four times!

Arctic Tern

This bird flies from northern parts of the globe all the way to the southernmost parts and back every year. This journey can be over 32,000 km (20,000 mi) long. Over the course of a tern's lifetime, it flies far enough to travel to the moon and back three times! The Arctic tern can travel very long distances without eating or sleeping. When it migrates, it searches for the perfect place to lay eggs and raise chicks.

Christmas Island Red Crab

These crabs live on a small Australian island off the coast of Indonesia. For two weeks in mid-November, tens of millions of crabs march together across the island's tropical forest. When they reach the coast, they lay their eggs.

Monarch Butterfly

These butterflies live in the United States during the summer. In the winter, they fly thousands of miles to northern Mexico. To survive this migration, they let the wind carry them whenever possible and use the sun's position to figure out which direction to fly. One hundred million butterflies make this journey every year. When they all take flight together, the sight is truly stunning.

Blue Pilchard

Billions of these small fish travel along the southern coast of Africa every year, searching for a safe place to lay their eggs. This mass migration attracts the attention of many predators, including dolphins, whales and sharks. But because they are swimming in such a large group, each fish is safer than if it were migrating alone.

Unexpected Friends

Sometimes one animal becomes friends with another type of animal. They might clean each other or help each other stay safe and find food. When the relationship helps at least one of the animals, this is called **symbiosis**.

Manta Ray and Cleaner Fish

In **coral reefs**, manta rays go to "cleaning stations" where different kinds of small fish take care of them. For example, fish called wrasses eat dead skin and other dirt on the rays' mouths and gills.

Giraffe and Oxpecker

Because they have such long necks, giraffes can't clean themselves very easily. Small birds called oxpeckers eat bugs and dirt off giraffes' bodies. In return for this help, the giraffes let the little birds ride around and even sleep on them. They live south of the Sahara Desert in Africa.

Clownfish and Anemone

Clownfish live inside anemones, which are bright underwater creatures that look like flowers. The clownfish scare away predators that might try to eat the anemones. In return, the anemones keep the clownfish safe with their stinging tentacles. The anemone's venomous stings don't hurt the clownfish but are painful for other creatures. They are found in the Pacific and Indian Oceans.

Sea Turtle and Suckerfish

The suckerfish, also known as the remora, is not very good at swimming! It doesn't have a swim bladder, an organ that helps most other fish swim. To move around, suckerfish ride on the backs of sea turtles and other animals. In exchange, they clean the animal's shell or skin, eating any leftover food they find along the way. Suckerfish and sea turtles live in warm ocean waters all over the world.

Nile Crocodile and Egyptian Plover

If you saw a little bird perched in this powerful predator's wide-open jaws, you might think the crocodile was about to have a snack! Actually, the plover is just doing some dental work. The bird uses its beak to clean old bits of food out of the crocodile's teeth, keeping the crocodile's mouth clean and healthy in exchange for a tasty meal. They live in Africa, south of the Sahara Desert.

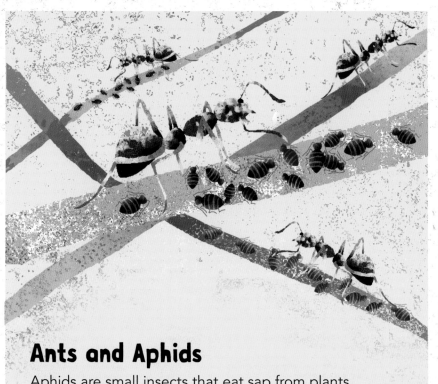

Ants and Aphids

Aphids are small insects that eat sap from plants. They produce a sugary liquid called honeydew as waste. Ants love to eat this honeydew, so they help protect aphids from predators. The ants also stroke the aphids with their **antennae** (also known as "feelers") to help them produce more honeydew. Both insects are found all over the world.

Expert Builders

Some animals find their homes in nature — and others build their own houses! These crafty creatures might use things they find in their environment as building blocks. Other animals make the materials they need.

Gardener Bird

These birds live in Australia and Papua New Guinea. Male gardener birds build extremely fancy homes to attract females. First they find a mossy spot at the foot of a tree where they weave a nest from twigs. Then they decorate it with objects like rocks, nuts, beetles, mushrooms and even man-made things like plastic bottle caps. The more decorations a nest has, the more attractive it is to a female.

North American Beaver

Beavers are the most impressive builders of any mammal. They use sticks to create their homes along rivers all over North America. A beaver house, called a dam or lodge, has an underwater entrance to guard against predators. Inside, beavers sleep, raise their young and stockpile food. Some massive beaver dams can be seen from satellites in outer space!

Sociable Weaver

Sociable weavers live together in large groups, building "nesting colonies" that they all share. They live in southern Africa, where their nests are easy to spot. Made of grass and twigs, they look like giant haystacks up in a tree. Sociable weavers can be found in groups of up to 300 birds, living in nests made of more than 100 small rooms. They will often help other birds in the group and even swap "apartments"!

Golden Orb-Weaver Spider

Golden orb-weaver spiders live in warm **climates** all over the world. As soon as the babies hatch out of their eggs, they start spinning an enormous circular web out of yellow (or "golden") silk. Their webs can hold up to 1 kg (2.2 lbs). That's strong enough to capture small birds, bats and lizards!

Grey Foam-Nest Tree Frog

This **amphibian** lives in the tropical swamps of central Africa. It builds a nest using its own sticky spit! Female frogs kick this liquid with their back legs, working it into a foam. They put the ball of foam in a tree, over a little pool of water. Then they lay their eggs on the foam. The nest hardens into a protective shell on the outside but stays soft and moist inside, keeping the eggs healthy and safe. After five days, the babies, called **tadpoles**, hatch and drop into the pool of water below.

Australian Compass Termite

These northern Australian insects build mud palaces, called mounds, that can hold three million termites! Termite mounds can be up to 3 m (almost 10 ft) tall. They are long and narrow, with one edge facing north and the other facing south, like a compass. The long, flat sides only face the sun in the morning and in the evening, when the sunlight isn't as strong. This helps the mound stay cool on the inside during the day.

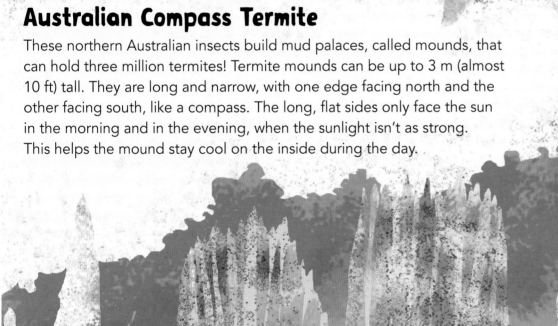

Devoted Parents

All parents have tough jobs and make sacrifices for their babies. But some animal mothers and fathers are truly superheroes!

Emperor Penguin

Female emperor penguins lay only one egg at a time. They place this egg on the father's feet. It is very cold where they live in Antarctica, so the father uses a special roll of fat at the bottom of his stomach to cover the egg and keep it warm. Then the mother goes off to the sea in search of food, while the father stays on land protecting the egg — for two months! The male penguins can't feed themselves or move during this time, so they huddle together to stay warm. When the females return, the chicks hatch and the fathers are finally free to move.

Italian Scorpion

The female Italian scorpion has up to 30 babies at a time. She gathers them onto her back to keep them safe. Then she carries them around everywhere with her for about a week, until they grow large enough to defend themselves. To avoid accidentally eating her own babies, the mother scorpion doesn't eat anything at all until her babies are able to protect themselves. The Italian scorpion is found in warm areas of Europe, Asia and Africa.

Seahorse

Mother seahorses lay up to a thousand eggs at a time, placing them in a special pouch on the father's stomach. The male seahorse keeps them safe there for up to 45 days. When they're ready to hatch, he squeezes the tiny baby seahorses out of his pouch in several exhausting bursts. Seahorses live in shallow, warm waters all over the world.

Bornean Orangutan

Orangutans give birth after a nine-month pregnancy, just like humans. But unlike human babies, orangutans stay in their mother's arms for the first two years of their lives! Since these orangutans live in trees on the island of Borneo, a baby orangutan must cling tightly to its mother's fur as she swings from one branch to another — sometimes more than 30 m (almost 100 ft) above the ground!

Horned Marsupial Frog

After this frog lays her eggs, they need to stay in a moist environment until they hatch. But horned marsupial frogs live in the rainforests of Central America, where there are many predators that would love to make a tasty meal of some frog's eggs. So the mother frog keeps them in a warm, damp pouch on her back. When they hatch, the mother frog places each baby frog on a plant leaf that collects little puddles of water.

American Alligator

A mother alligator's toothy jaws might look terrifying — but her mouth is a safe place for her babies! Alligators in the southern United States lay their eggs in a mud nest in the ground and bury them with dirt to keep them warm. When ready to hatch, the babies make a high-pitched barking sound. Their mother hears them and digs up the eggs to release them. Then she picks up her babies gently in her mouth and brings them into the water.

Special Snoozers

Everyone needs sleep to stay healthy. Some animals snooze for a long time all at once, while others sneak in little naps throughout the day.

Koala

The koala sleeps up to 18 hours a day! It lives in western Australia, where it spends most of its time in a type of tree called a eucalyptus. A koala can eat almost 1 kg (2.2 lbs) of eucalyptus leaves a day. Then it spends the rest of its time relaxing while it digests all that food!

Giraffe

The giraffe only sleeps for two hours a day, divided up into lots of short little naps. Giraffes often sleep standing up, because they are safer that way. The African grasslands where they live do not offer many places to hide. When they lie down, their long necks make them a good target for predators, such as lions. So if they do lie down on the ground, they only nap for a very short time — six minutes at most.

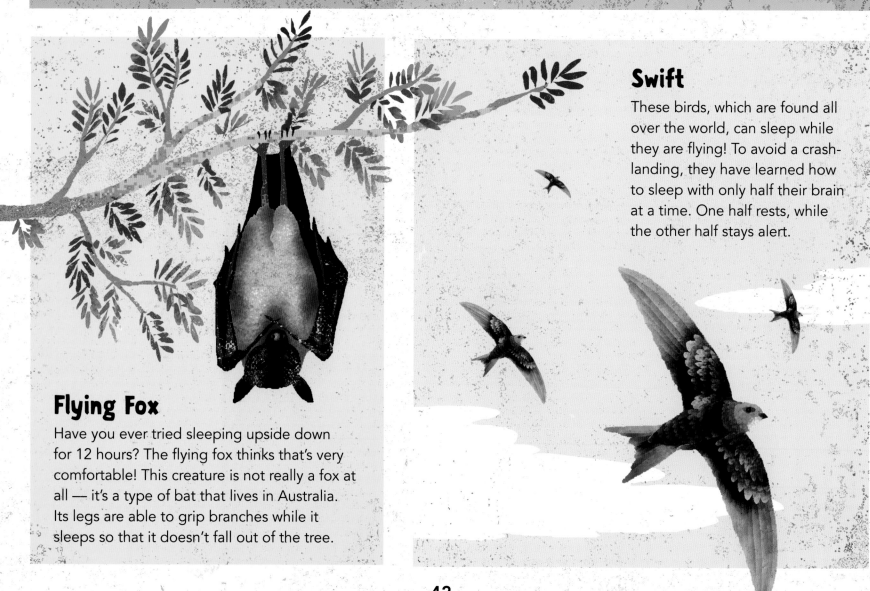

Swift

These birds, which are found all over the world, can sleep while they are flying! To avoid a crash-landing, they have learned how to sleep with only half their brain at a time. One half rests, while the other half stays alert.

Flying Fox

Have you ever tried sleeping upside down for 12 hours? The flying fox thinks that's very comfortable! This creature is not really a fox at all — it's a type of bat that lives in Australia. Its legs are able to grip branches while it sleeps so that it doesn't fall out of the tree.

Zebra Finch

Scientists believe that this Australian bird may have musical dreams! They have observed the zebra finch's brain while it is singing and discovered that it uses those same areas while sleeping.

Parrotfish

The parrotfish lives in warm, shallow coral reefs. To protect itself at night, this fish rolls itself up in a ball of **mucus** while it sleeps. It makes this slimy substance in its mouth and then spits it out. The mucus has a disgusting smell, which hides the parrotfish's real scent and keeps predators away.

Hippopotamus

During the day, this enormous African mammal rests in pools of water to stay cool. While it's sleeping, it sinks down towards the bottom of the water, then floats back up so it can take a few breaths of air. It does this over and over again without waking up!

Who Glows There?

Some animals have **bioluminescence**, or the ability to make light with their bodies. Glowing might help them find prey, attract a mate or hide from predators. Many of these animals live in deep, dark parts of the ocean, but some live on land as well.

Viperfish

The viperfish is one of the fiercest — and strangest! — predators in the deepest parts of the Pacific Ocean. It has about 350 tiny light-producing organs along the sides of its body. When it turns on this glow, smaller fish become curious about the bright light and will swim right between its teeth to have a look. Then it's easy for the viperfish to gobble them right up!

Motyxia Millipede

These insects only live in the Sierra Nevada mountains of the western United States. Their green-blue glow is a warning to predators. When the millipedes feel threatened, they ooze a powerful poison.

Velvet Belly Lantern Shark

This is one of the smallest sharks in the world. It only grows to be about as long as a person's arm. It lives in the northeastern Atlantic Ocean, where its glowing belly helps it find potential mates and also distracts predators.

Hawaiian Bobtail Squid

This squid breathes in tiny bioluminescent **bacteria** that make its body glow. It lives in shallow waters near Hawaii, where it buries itself in the sand during the day. The glowing bacteria also help it cast shadows to confuse predators at night when it comes out to hunt.

Glowworm

Glowworms are not actually worms — they're beetles! Females give off a greenish-orange glow at night. Their light attracts males, which have very light-sensitive eyes. Some types of glowworms also have a red light on their heads that helps them find prey and scare off predators. They live in Europe and the Americas.

Firefly

At the beginning of the summer, if you walk past a field or forest in the evening, you might see a beautiful sight: tiny floating yellow-green lights. These are fireflies! The flying insects have blinking lights that help attract mates. Every type of firefly has its own special pattern of blinks and flashes. This helps different types of fireflies to find each other. These insects are found all over the world.

Weird and Wacky

In this section, you'll find some unusual
animals you might not have seen before!
Each one has a unique feature, whether
it's a clever way of finding a mate or
a strange nose unlike any other.

Frigate Bird

Male frigate birds have a bright red pouch under their beaks that they can blow up like a balloon. When they are looking for a mate, they put on quite a performance! They puff this pouch full of air, spread their wings wide, point their beaks to the sky and make a loud drumming sound. Frigate birds live near warm oceans all around the world.

Red-Lipped Batfish

The red-lipped batfish looks like it is wearing lipstick. Even though it lives in the water, this bizarre creature can also use its fins like legs to walk on the bottom of the ocean! It is only found in the Pacific Ocean near the Galapagos Islands.

Thorny Devil

This little reptile might look sharp and scary, but it's really quite harmless. Its spiky skin scares away predators. The spines on its back also help it survive in the hot, dry deserts of central Australia. Dew collects in between these spines. Then the drops of water roll through grooves on the lizard's skin towards its mouth so it can have a drink.

Aye-Aye Lemur

This primate is the size of a cat and has teeth like a rodent's. But its most unique features are its fingers. It uses one finger to make a tapping sound that helps it find other lemurs. Another of its fingers is very long with a hooked nail that lets it dig insects out of trees to eat. The aye-aye lives in Madagascar.

Pink Fairy Armadillo

The pink fairy armadillo is nocturnal and spends most of its time hiding underground. It lives in the desert in central Argentina, where it digs burrows in the sand using its large, strong claws. It is nicknamed the "sand swimmer" because it can tunnel so quickly it looks almost like a fish swimming in the sea. Its unique pink shade comes from blood vessels under its shell, which turns pinker when it needs to cool down and paler when it's trying to warm itself up.

Saiga Antelope

This endangered antelope lives in central Asia. It has an incredibly long, flexible nose. Scientists think that this special nose helps filter out dust when antelope herds are migrating across long distances and kicking lots of sand into the air.

Star-Nosed Mole

This little North American mole has a bizarre nose with 22 tentacles that can move by themselves. This special nose is only the size of a human fingertip, but it helps the mole have an incredible sense of smell. The mole's nose helps it sniff out insects to eat, since it cannot see very well.

Dumbo Octopus

This small octopus lives in very cold, deep ocean waters around the world. It has two big fins on the sides of its head that look like elephant ears — which is where it gets its name! These special fins suck in and push out water, which helps the octopus change direction when it swims.

Bold and Bright

Some animals show off their vibrant feathers, fur, skin or scales to attract a mate. Others use their flashy appearance to warn predators that they are poisonous . . . or just to scare them away!

Chameleon

The chameleon is famous for being able to blend in with its surroundings. But sometimes, this reptile will also change the patterns on its skin to frighten a predator or attract a mate. Chameleons are found in warm areas around the world. If the weather is very hot, the chameleon can make its skin paler to reflect sunlight, which helps it stay cool and comfortable. Chameleons can also move their eyes separately so they can look in two different directions at once!

Red-Eyed Tree Frog

This nocturnal amphibian lives in the rainforests of Central America. It is not poisonous, but its bright red eyes look scary to predators, helping to keep the frog safe.

Nudibranch

This invertebrate is also sometimes called a "sea slug." Nudibranchs can be almost any shade of the rainbow and are found in oceans all over the world. They absorb and store venom from the anemones and coral that they eat. Their bright skin warns predators that they can use this poison to protect themselves.

Mandrill

The mandrill, which lives in western Africa, is the largest monkey in the world. The males have bright red and blue noses and golden fur around their necks. Females are attracted to the brightest and largest males.

Lilac-Breasted Roller

The beautiful lilac-breasted roller lives in southern Africa. It is the national bird of both Kenya and Botswana. Its long tail helps it perform acrobatic twists, loops and dives in midair while flying.

Wattle Cup Caterpillar

The bright appearance of this Australian caterpillar is a warning sign. The wattle cup caterpillar's fierce-looking spikes deliver a sting stronger than three wasps! Surprisingly, this caterpillar turns into a very plain-looking white moth when it goes through **metamorphosis**.

Ruby-Throated Hummingbird

This tiny bird lives in North America, where the strong sunshine makes the male's red neck feathers stand out. Unlike most other birds, it can stop and hover while feeding in midair.

Mandarin Duck

The Mandarin duck lives in Asia, where the male's rainbow of feathers help him attract a mate. The female's feathers, on the other hand, allow her to blend in with her surroundings while sitting on a nest of eggs. This helps her keep her eggs safe from predators.

Mandarin Fish

The Mandarin fish lives in coral reefs in the Pacific Ocean. It uses its bright scales to attract mates and warn away predators. It has two special fins that let it "walk" along the sea floor, where it finds fish eggs and small creatures to eat.

Strange Superpowers

Some of these animals have figured out how to survive in places where almost nothing else can live. Others have found ways to avoid aging, sickness or even the need to breathe!

Reindeer

Only a few creatures can survive at the North Pole, where the temperatures can reach -50°C (-58°F). Reindeer's yellow eyes turn blue in the winter, which helps them capture more light during the darker months of the year. Their noses heat the air before they breathe it into their lungs. And their hooves, which have spongy, soft pads in the summer, turn harder in the winter so that they can break through ice.

Immortal Jellyfish

This tiny jellyfish, only the size of your smallest fingernail, seems to have discovered the secret to living forever! It starts life as a **polyp**, a creature with a tube-shaped body attached to the sea floor. Then it develops into an adult jellyfish that can swim free. But if it gets injured, it can turn back into a polyp and heal itself. Then it changes back into an adult and swims free again. Unless it gets eaten by another animal, it seems to be able to repeat this cycle endlessly! The immortal jellyfish is found in tropical waters around the world.

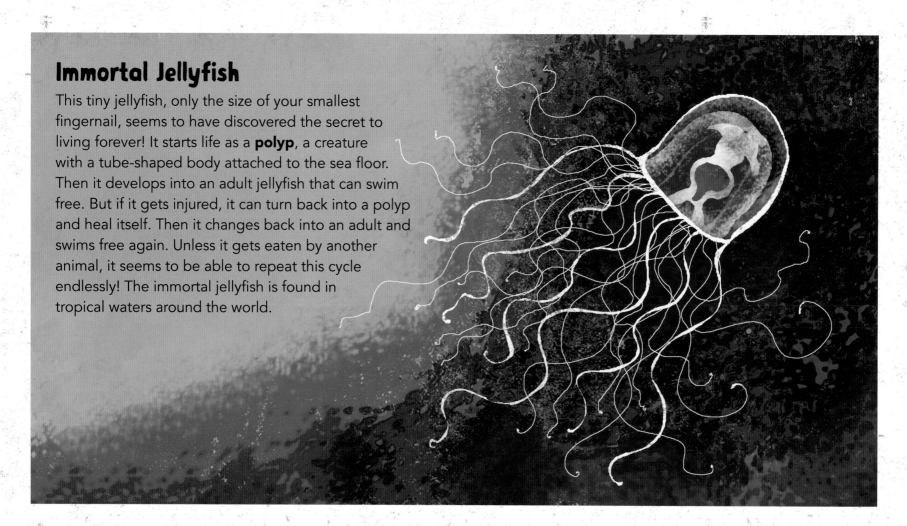

Water Bear

The water bear is only about the size of a grain of sand! They can live almost anywhere, from rainforests to deep oceans, from boiling water to Antarctic ice. The water bear can survive for more than ten years without any water and can even live in outer space. They seem to make special chemicals in their bodies that protect them from cold, heat and drying out. But scientists still do not quite understand how this works!

Olm

The olm is a type of salamander, an amphibian that looks like a lizard. It lives in water-filled underground caves in Italy and the Balkan mountains. There is very little light where it lives, so the olm's eyes never fully develop. It uses its senses of smell and hearing to find food. But because it is also hard for the olm to find enough to eat, its superpower is that it can go without food for about ten years!

Wood Frog

This North American amphibian has figured out a superpower for handling extreme cold. Instead of using up energy trying to stay warm in the winter, the wood frog can let parts of its body freeze into ice! Its breathing, heartbeat and blood flow also slow down until the weather warms up and the frog can thaw.

Naked Mole-Rat

Naked mole-rats get their name from the fact that they have no fur. They live in eastern Africa, where they dig complicated underground burrows and live in large groups. Naked mole-rats are able to survive for almost 20 minutes without breathing any oxygen!

Fennec Fox

This fox lives in the hot, dry deserts of Africa and the Middle East. It can manage for long periods of time without drinking any water. Instead, it gets the water it needs from eating fruit, leaves and plant roots. It also has enormous ears, which help keep its body cool and can even hear prey moving underground.

Animal Groups

Animals are a category of living creatures. The six main types of animals are mammals, invertebrates, birds, reptiles, amphibians and fish. Here are all the animals in this book, sorted by type.

Mammals give birth to live young, rather than laying eggs. They also make milk to feed their babies.

Cheetah

Orca

Wolverine

Wolf

Chimpanzee

Bottlenose dolphin

African elephant

Leopard

Gnu

Grey whale

Giraffe

North American beaver

Bornean orangutan

Koala

Flying fox

Hippopotamus

Aye-aye lemur

Pink fairy armadillo

Saiga antelope

Star-nosed mole

Mandrill

Reindeer

Naked mole-rat

Fennec fox

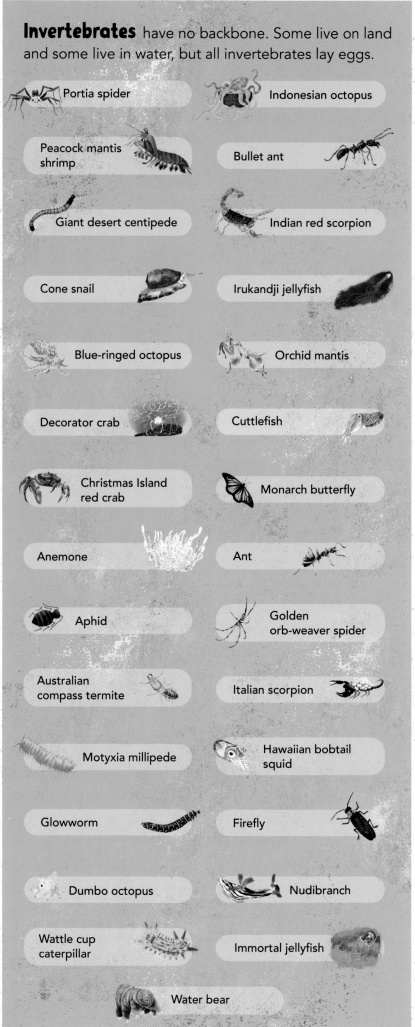

Invertebrates have no backbone. Some live on land and some live in water, but all invertebrates lay eggs.

Portia spider

Indonesian octopus

Peacock mantis shrimp

Bullet ant

Giant desert centipede

Indian red scorpion

Cone snail

Irukandji jellyfish

Blue-ringed octopus

Orchid mantis

Decorator crab

Cuttlefish

Christmas Island red crab

Monarch butterfly

Anemone

Ant

Aphid

Golden orb-weaver spider

Australian compass termite

Italian scorpion

Motyxia millipede

Hawaiian bobtail squid

Glowworm

Firefly

Dumbo octopus

Nudibranch

Wattle cup caterpillar

Immortal jellyfish

Water bear

Birds have wings, feathers and a beak. Most birds lay eggs in nests and can fly.

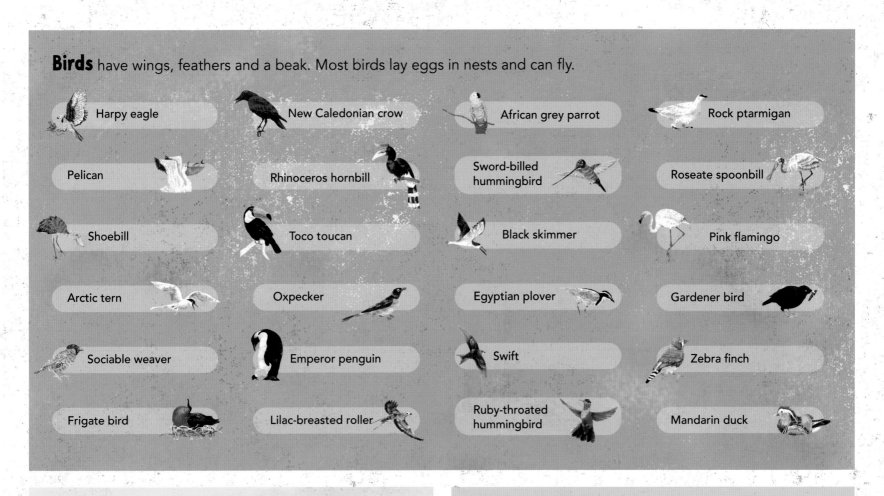

Harpy eagle	New Caledonian crow
African grey parrot	Rock ptarmigan
Pelican	Rhinoceros hornbill
Sword-billed hummingbird	Roseate spoonbill
Shoebill	Toco toucan
Black skimmer	Pink flamingo
Arctic tern	Oxpecker
Egyptian plover	Gardener bird
Sociable weaver	Emperor penguin
Swift	Zebra finch
Frigate bird	Lilac-breasted roller
Ruby-throated hummingbird	Mandarin duck

Reptiles all have dry, scaly skin. They lay their eggs, which have soft, leathery shells, on land.

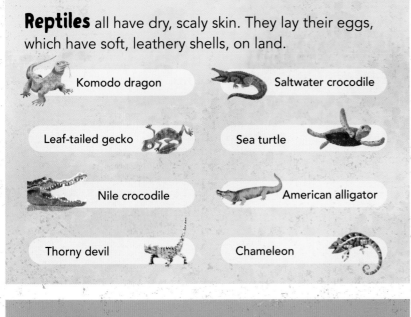

Komodo dragon	Saltwater crocodile
Leaf-tailed gecko	Sea turtle
Nile crocodile	American alligator
Thorny devil	Chameleon

Amphibians are born in water and breathe through gills like fish, but later develop lungs so that they can live on land as adults. They lay their eggs in water.

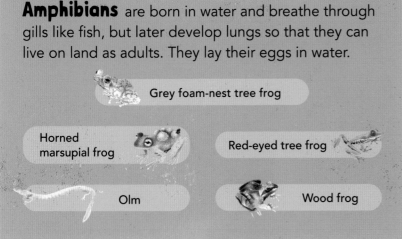

Grey foam-nest tree frog	
Horned marsupial frog	Red-eyed tree frog
Olm	Wood frog

Fish are animals with gills, fins and scales. Most fish lay eggs and spend their lives swimming in water.

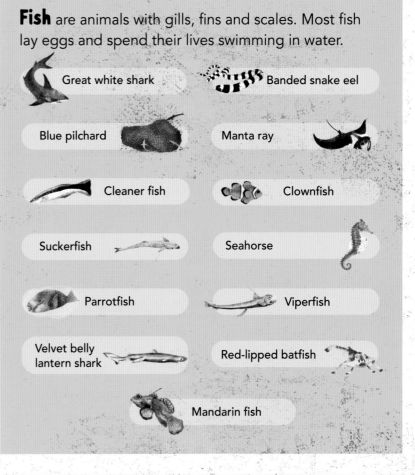

Great white shark	Banded snake eel
Blue pilchard	Manta ray
Cleaner fish	Clownfish
Suckerfish	Seahorse
Parrotfish	Viperfish
Velvet belly lantern shark	Red-lipped batfish
Mandarin fish	

What other animals do you know in each of these categories?

Glossary

algae: a type of simple, green water plant

alpha or apex predator: an animal that is not hunted by any other animals

amphibian: an animal that is born in water but lives on land as an adult. Some examples are frogs and toads.

anemone: a bright, venomous sea creature that looks like a flower with stinging arm-like parts called tentacles

antennae: long, thin sensors (also called "feelers") attached to the heads of some insects

arachnid: a type of creature with eight jointed legs and a two-part body

bacteria: tiny creatures made of only one cell

bioluminescence: the ability of some animals to make their own light or glow

camouflage: an animal's ability to blend in with its surroundings, sometimes by changing the way it looks

cell: the smallest unit of a living thing

climate: the usual weather conditions of a specific area

coral: a collection of tiny polyps (tube-shaped sea creatures) that cannot move and leave behind a hardened, plant-like skeleton when they die

coral reef: a warm-water ocean environment made up of coral skeletons, where many plants and animals live

crustacean: a type of invertebrate that has a hard shell and usually lives in the water; includes crabs, lobsters and shrimp

evolve: for an animal species to change gradually over generations, adapting to their environment in order to better survive

exoskeleton: a hard, protective outer shell

food web: a diagram that shows a set of living things connected by the animals or plants that they eat

habitat: the place where an animal lives

invertebrate: an animal with no backbone. Some examples are insects and jellyfish.

keratin: a tough material that is an important part of hair, feathers, hooves, claws and fingernails

mammal: an animal that gives birth to live young. Along with many of the animals in this book, people are also mammals!

matriarch: the female leader of a group or family

metamorphosis: the process in which an insect or amphibian's body transforms as it enters a new stage of its life, such as a caterpillar turning into a moth or butterfly, or a tadpole turning into a frog

migrate: to move from one habitat to another, for part of the year or part of an animal's life. A mass migration is the movement of many animals together.

mucus: a slimy, wet substance that some animals (like the parrotfish) produce for protection

nocturnal: only active at night

polyp: a type of tube-shaped creature that is attached to the ocean floor and has a ring of tentacles around its mouth

predator: an animal that eats other animals

prey: an animal that gets eaten by another animal

primate: a type of mammal that has a highly developed brain and can grab things with its hands; includes monkeys, apes and humans

proboscis: a long tube attached to or near an animal's mouth that it uses to eat

rainforest: a warm woodland area that gets at least 254 cm (100 in) of rain a year

reptile: an animal with dry, scaly skin that lays soft-shelled eggs. Some examples are lizards and turtles.

species: a group of living things that all share something in common and can reproduce together

symbiosis: a relationship between two creatures that both help each other in some way

tadpole: the early life stage of a frog, toad, newt or salamander

talon: a predator bird's sharp claw, used for hunting prey

tentacle: a long, flexible body part that has suckers at the tip. An octopus has arms, not tentacles, because they have suckers all along their length.

tundra: a cold, northern area where the ground stays frozen all year long

venom: a poisonous liquid that some animals (like octopuses, scorpions and anemones) produce to defend themselves or attack prey

wingspan: the distance across a bird's outstretched wings, from the tip of one wing to the other

Barefoot Books
2067 Massachusetts Ave
Cambridge, MA 02140

Barefoot Books
29/30 Fitzroy Square
London, W1T 6LQ

Original Italian edition: © Dalco Edizioni Srl
Via Mazzini 6 – 43121 Parma, Italy
www.dalcoedizioni.it
Text by Dunia Rahwan
Illustrations by Paola Formica

Translation copyright © 2020 by Barefoot Books
The moral rights of Dunia Rahwan and
Paola Formica have been asserted

First published in United States of America by
Barefoot Books, Inc and in Great Britain by
Barefoot Books, Ltd in 2020
All rights reserved

Graphic design by Sarah Soldano and
Lindsey Leigh, Barefoot Books
Edited by Lisa Rosinsky, Barefoot Books
Translation support provided by Danielle Buonaiuto
Reproduction by Bright Arts, Hong Kong

Printed in China on 100% acid-free paper
This book was typeset in Avenir and Graphen
The illustrations were prepared using digital painting

Hardback ISBN 978-1-64686-066-1
E-book ISBN 978-1-64686-083-8

British Cataloguing-in-Publication Data:
a catalogue record for this book is
available from the British Library

Library of Congress Cataloging-in-Publication Data
is available under LCCN 2020011976 (print)
and LCCN 2020011977 (e-book)

3 5 7 9 8 6 4 2

Barefoot Books would like to
thank **zoologists Laura Isaacs
and Kimberly Warren** for their
help in the editing of this book.

Dunia Rahwan studied biology and is now a
science journalist, making regular contributions to
periodicals for adults and children. Passionate about
nature photography, scuba diving, animals and
travel, she lives in Milan with her cat and two dogs.

Paola Formica has illustrated novels and picture
books as well as periodicals and textbooks. Her
artwork has won numerous international awards.
She lives in Milan with her family and their two cats,
three turtles and a tank full of tropical fish.